AF811279

THE

# Meet and Greet

PLAYBOOK

How to Make Powerful First Impressions with

Customers, Clients, and Guests

Bruce Huddleston

The Car Sales Survival Series

Bedrock Heritage Publishing

**The Meet and Greet Playbook**

How to Make Powerful First Impressions with Customers, Clients, and Guests



Published by:

Bedrock Heritage Publishing

A Division of Life Guidance Consulting LLC

Tyler, Texas

www.bedrockheritagepublishing.com

info@bedrockheritagepublishing.com

ISBN: 978-1-972179-07-9

ISBN: 978-1-972179-45-1 (EPUB)

Part of The Car Sales Survival Series

First Edition, 2026

Printed in the United States of America

# DISCLAIMER

This book is based on the author's personal and professional experiences, observations, and opinions, accumulated over a 35-year career in the automotive industry. It is intended for educational and informational purposes only.

The stories and anecdotes contained in this book are drawn from real-world situations encountered throughout the author's career. However, names, identifying details, specific circumstances, employer names, dealership names, and individual characteristics have been changed, omitted, combined, or fictionalized to protect the privacy of the individuals involved. Any resemblance to specific living persons, current or former employers, or existing businesses is coincidental and unintentional.

No individual, dealership, organization, or employer referenced or implied in the stories within this book has reviewed, approved, or endorsed the content herein. The recollections and characterizations presented are solely the author's own perspective and memory of events and do not constitute a factual record, legal testimony, or statement of fact regarding any identifiable person or entity.

The sales strategies, techniques, and professional advice presented in this book reflect the author's personal approach and experience. Individual results will vary based on experience, effort, market conditions, dealership policies, and other factors beyond the author's control. Nothing in this book constitutes a guarantee of income, employment, or professional outcome.

The author and publisher have made reasonable efforts to ensure the accuracy of information presented at the time of writing. The author and publisher make no representations or warranties regarding the completeness, accuracy, or current applicability of the information contained herein, and expressly disclaim any liability arising from the use or application of the content of this book.

By reading this book, you acknowledge and agree that the author and publisher shall not be liable for any damages, losses, or claims arising directly or indirectly from the use of or reliance upon any information contained herein.

To the professionals who understand that a simple greeting

can change the entire customer experience.

And to every salesperson who ever lost a customer in the first ten seconds

and had to figure out why the hard way.

# FREE BONUS FOR READERS

## Your Complete Digital Script Library

As a reader of The Meet and Greet Playbook, you have access to the complete digital version of every script and checklist in this book — formatted, printable, and ready to customize in your own voice.

Visit the link below, enter your email address, and receive:

**The Complete Meet and Greet Script Library**

All scripts from this book are in a digital, editable format

**The Full Meet and Greet Checklist**

Printable and ready for daily use

**Additional Objection Handling Scenarios**

Expanded responses for the toughest customer moments

**Access to the Car Sales Survival Series**

Browse and order all titles from the series.

www.carsalessurvivalseries.com/scripts

Enter your email to claim your free reader bonus.

# Contents

INTRODUCTION

# Why the Greeting Is the Most Underrated Skill in Business

Most people think a meet-and-greet is simple. Say hello, introduce yourself, and ask how you can help. On the surface, that sounds right. But anyone who has worked with customers for any length of time knows the reality is more complicated than that.

The meet and greet is not only about the words that come out of your mouth. It is about the impression you create the moment a customer sees you. It is about whether people feel welcome or whether they feel like they just interrupted someone's day. It is about whether they leave that first interaction with their guard up or their guard down. And it starts — every single time — before you open your mouth.

Before a customer hears your product knowledge, your presentation, or your price, they are already deciding how they feel about you. They notice how you carry yourself. They notice whether you look alert or distracted. They notice whether your face says "welcome" or "why are you here." Within seconds, they're asking themselves a question they may never say out loud: Can I trust this person to help me?

That silent question is why the meet-and-greet matters more than most people realize. The first few moments of an interaction often determine everything that follows. Get the greeting right, and the customer relaxes. Get it wrong, and you're spending the rest of the conversation working uphill, trying to recover trust you never should have lost.

I've been in the car business for thirty-five years. I've trained hundreds of salespeople. And the single most consistent thing I've seen separate the ones who build careers from the ones who burn out is this: the professionals who last understand that the greeting is not a formality. It is the foundation of everything that comes after it.

This book is a training guide, not a theory lecture. It is written for salespeople, retail associates, front desk staff, managers, and anyone else who works face-to-face with customers and wants to get better at the moment that matters most. The principles apply across industries because people are still people. Everyone wants to feel acknowledged, respected, and helped. Nobody wants to feel like a burden.

What you'll find in these pages is what I've seen work on the floor, in the training room, and over thirty-five years of real interactions with real customers. Some of those stories are funny. Some of them cost me money before I understood the lesson. All of them are real because that's how this stuff actually gets learned.

Read it straight through the first time. Then use it as a reference. Come back to the checklist before a shift. Come back to the stories when you need a reminder of what not to do. Come back to the psychology chapter when a customer you can't figure out is giving you trouble.

The goal is simple: help you make the first moment count. Because in this business, first moments are the ones you can't afford to waste.

---

**"The Rule: The greeting is not where the sale is won. It is where the sale is lost. Everything else depends on getting this right."**

---

CHAPTER ONE

# The Meet and Greet Starts Before You Speak

The Impression That Forms Before the Conversation Begins

One of the most common misunderstandings in sales and customer service is the idea that the meet-and-greet begins when you start talking. It doesn't. It begins the moment the customer sees you.

That moment might happen when they pull onto the lot and look toward the showroom. It might happen when they walk through the store's front door and scan the room. It might happen when they approach a front desk, and someone is standing behind it. Wherever and whenever they first lay eyes on you, the meet and greet has already started — and you haven't said a word yet.

Customers observe more than most employees realize. They notice whether you're standing upright or slouched against a wall. They notice whether you look alert or bored. They notice if you're on your phone, if you're deep in a conversation with a coworker, or if you look like helping them is going to be an inconvenience. All of that communicates something — before you've had the chance to say hello.

## You Are Already Communicating

Here's the uncomfortable truth: you don't get to choose whether you're making a first impression. You only get to choose what kind. The customer is reading you from the moment they notice you, whether you're ready or not.

Posture communicates confidence or its absence. A salesperson standing upright, head up, aware of their surroundings — that person looks ready to help. A salesperson slouched in a chair scrolling through their phone — that person communicates that this job is something they're enduring, not something they take seriously. The customer sees both of those things instantly.

Your energy communicates, too. There's a version of this business where you walk out onto the floor genuinely glad to be there, ready for whoever comes through the door. And there's a version where you're running on empty from the last bad interaction and haven't reset yet. Customers feel the difference. They may not be able to articulate it, but they feel it — and they respond to it.

Appearance matters as well. This doesn't mean expensive clothes or a particular look. It means appropriate and professional for the environment you're in. A well-kept appearance communicates that you take the job seriously. It tells the customer that they're walking into a professional environment, not a casual one. And it gives them one less reason to put their guard up before you've even opened your mouth.

## Be Ready Before They Arrive

The practical implication of all this is simple: your preparation has to happen before the customer gets there, not after. You can't clean up your posture after they've already seen you slumped. You can't put your phone away after they've already watched you scroll for thirty seconds. Those impressions are formed in real time, and once they're formed, you're working uphill.

This means being presentable before your shift starts. It means staying attentive even when traffic is slow and nothing is happening. It means putting distractions away and keeping your attention on the floor, the lot, or the entrance — wherever your customers come from. It means understanding that the job starts the moment someone might be watching, not the moment you decide to engage.

The professionals who understand this don't leave the first impression to chance. They make a deliberate choice about what they're communicating before anyone walks through the door. And that choice — to be ready, to be present, to look like someone worth talking to — is one of the simplest and most powerful things you can do for your numbers.

---

**"The Rule: The meet and greet starts the moment they see you. Be ready before they arrive, because once they do, the clock is already running."**

---

CHAPTER TWO

# What You Communicate Before You Say a Word

Silent Signals That Build Trust or Destroy It

Before a single word is spoken, a customer has already started forming an opinion. Not consciously, not deliberately — but automatically, the way human beings are wired. They're scanning for signals that tell them whether this environment is safe, whether this person is trustworthy, and whether it's worth their time to keep going. Those signals are almost entirely nonverbal.

Understanding what you're communicating before you speak is not a soft skill. It is a practical one. Every element of your nonverbal presentation either builds the customer's comfort or erodes it. There is no neutral.

## Posture

Posture is the first thing a customer reads from a distance. Standing upright with relaxed shoulders communicates confidence and readiness. It says: I'm here, I'm present, and I'm capable of helping you. Slouching, leaning, or appearing physically collapsed into whatever surface is nearby says the opposite — even if that's not your intention.

It doesn't require military bearing. It requires awareness. Know how you're standing. Make a habit of checking it. A small adjustment in how you carry yourself can change the entire first impression before you've done anything else.

## Eye Contact

Natural eye contact shows the customer you have their attention. It communicates focus and honesty. It says: " You are what I'm looking at right now, and nothing else is competing for my focus.

That doesn't mean staring. Intense, unbroken eye contact creates a different kind of discomfort. What you're going for is calm, attentive awareness—the kind of eye contact that makes a person feel seen without feeling studied. When you're glancing at your phone, looking past the customer at something across the room, or scanning the environment while they're trying to talk to you, the message you're sending is that they're not the priority. Customers notice that immediately.

## Facial Expression

Your face sets the emotional tone of the interaction before you say a word. A relaxed, open expression — even a small genuine smile — lowers tension. It tells the customer that their presence is welcome. A flat expression, a furrowed brow, or the blank look of someone who's been on their feet for six hours and is running on autopilot creates distance.

You cannot fake this indefinitely. That's why what I call the Disneyland Principle matters: you have to intentionally decide to be fully on, present, and ready to step on stage every time you interact with people.

You have to make the choice, before the customer gets there, that you're going to show up with the right energy. If you're carrying frustration from the last interaction — a deal that fell apart, a rude customer, a manager who made your morning harder — you have to reset before the next person gets a version of that.

## Movement and Pace

How you move communicates as much as how you stand. Quick, jerky movement makes people feel like something is wrong — like you're anxious, rushed, or chasing them down. A calm, deliberate pace communicates the opposite. It says: I'm in control of this environment, and there's no reason to be tense.

This applies to the approach more than anywhere else. The speed at which you walk toward a customer tells them a great deal about what they're in for. Too fast, and they feel hunted. Too slow, and they feel ignored. The right pace is confident and purposeful — the pace of someone who is glad a customer arrived and is moving to help them, not sprinting to claim them.

## What You're Holding

This one gets overlooked, but it matters. Walking up to a customer while holding your phone — even if it's in your hand and not in front of your face — communicates that the phone might pull your attention at any moment. It introduces a subtle but real sense of competition for your focus. The same is true of carrying unrelated paperwork, a coffee cup in the middle of taking a drink, or anything else that signals you were in the middle of something else before they interrupted.

When you're about to greet a customer, your hands should be empty and relaxed. That's a small thing that communicates something large: right now, the only thing that matters is them.

---

**"The Rule: You are always communicating. The only question is whether you're doing it on purpose."**

---

CHAPTER THREE

# The Acknowledgment

The First Thing You Do — Before You Say Anything Else

There are two ways to fail a customer in the first thirty seconds without saying a single wrong word. The first is to ignore them. The second is to rush them. Both communicate the same thing from the customer's perspective: this person is either not paying attention or is desperate. Neither one inspires confidence.

The acknowledgment is the solution to both. It is the first active step you take during the meet-and-greet, and it costs almost nothing. A nod of the head. A small wave. A brief "Hi folks, I'll be right with you" from a comfortable distance. That small gesture does three things at once: it tells the customer they've been seen, it tells them help is available, and it gives them a moment to settle in without feeling ambushed.

Most customers arrive slightly guarded. They've heard the stories — the pushy salesperson, the pressure tactics, the experience they were warned about before they walked in. A calm acknowledgment from a distance does something quietly powerful: it begins to lower that guard before the conversation has even started. You've already exceeded a low expectation by doing almost nothing.

## Timing Matters

The acknowledgment should happen quickly — within moments of the customer appearing — but not with urgency. There's a difference between prompt and frantic. Prompt says: "I noticed you, and I'm available." Frantic says: I've been waiting for someone to walk through that door, and I'm not letting you get away.

If you're with another customer or finishing something that genuinely requires your attention, a simple acknowledgment buys you the time to finish properly. "Welcome in — I'll be with you in just a moment" is not a brush-off. It is a professional communication that respects both the customer waiting and the one you're currently serving. Customers will wait if they feel acknowledged. They won't wait if they feel invisible.

## What Happens When You Skip It

I've watched more deals die in the acknowledgment phase than in any other part of the process — including the close. Not because anything dramatic happened, but because nothing happened. The customer walked in, looked around, felt like no one noticed them, and their comfort level dropped. By the time someone finally approached, the customer was already slightly irritated. That irritation had to be overcome before any real conversation could begin.

Every minute a customer stands in your environment feeling unacknowledged is a minute they're deciding whether to stay. The longer that goes on, the more likely they are to decide they don't. A simple acknowledgment stops that clock immediately.

### FROM THE FLOOR

My wife and I pulled onto a dealership lot on a hot afternoon — ninety-five degrees, which is nothing unusual in Texas. We got out and started walking the inventory. Three salespeople were standing on an elevated deck watching us from a distance.

Nobody moved.

They just stood there and watched while we walked around their lot looking at vehicles. I am not going to walk up to a platform and beg someone to help me spend money. So we got back in the car and drove to the dealership next door. Bought a car there the same afternoon.

The salespeople on that deck never said a wrong word. They never pressured us, never used a bad line, never made a mistake in the conversation. Because there was no conversation, they lost a sale before it ever started—not by doing something wrong, but by doing nothing at all.

That's the cost of a missed acknowledgment. It's not always dramatic. Sometimes it's just a customer who drives away while you're still standing on your deck, wondering why nobody's buying today.

---

**"The Rule: Acknowledge every customer the moment they arrive. Not with pressure — with presence. Let them know they've been seen before you've said a word."**

---

CHAPTER FOUR

# The Approach

How You Walk Toward a Customer Tells Them Everything

The acknowledgment got their attention. Now you move. And how you move — the pace, posture, and body language of the approach itself — communicates something to the customer with every step.

This is the moment most training programs underestimate. A lot of attention gets paid to what you say in the greeting, and almost none to what happens in the four or five seconds before you get there. But customers are watching those seconds closely. What they see shapes how they receive everything that comes after.

## The Pace of the Approach

Walk toward the customer at a normal, purposeful pace. Not rushing, not dragging. The pace of someone who is genuinely glad a customer arrived and is moving to help them — not the pace of someone who's been sitting on their heels all day and is making sure this one doesn't get away.

The sprint approach is one of the most common mistakes in the business, and it happens for understandable reasons. You see a customer, and you don't want to lose the up. Or you've been slow all day, and the arrival of someone on the lot feels like a lifeline. So you move fast. You get there quickly. And the customer, who just stepped out of their car and hasn't even had a chance to look around yet, immediately feels like they're being chased.

That feeling — of being chased, of being descended upon before you're ready — is one of the things customers dread most about this environment. When you trigger it in the first five seconds, you've already set the wrong tone for everything that follows. Speed communicates desperation. And desperation is the last thing any customer wants to feel coming from the person who's supposed to help them.

## Hands Visible, Head Up

As you approach, keep your hands visible and relaxed at your sides or in front of you in a natural position. This is not a formal requirement—it's a trust signal. Visible hands communicate openness. They tell the customer, at a subconscious level, that there is nothing to be guarded about.

Head up, eyes forward. You're moving toward them, not looking at the ground as you walk. Eye contact from a reasonable distance, even before you've reached them, says: I see you, I'm coming to help you, and I'm confident in what I'm doing.

### FROM THE FLOOR

I was sitting in the front lobby of a dealership I managed — a glass-front wall with a full view of the lot. A car pulled in. Two salespeople inside the showroom both saw it at the same time. They jumped up, ran for the door, and literally shoved each other trying to get through it first — pushing and shoving like it was a race.

The customers, still in their car, watched every second of this.

The one who won was out of breath when he got there. He stuck his hand out and started shaking the man's hand before he'd even introduced himself, and never looked at the wife. Never acknowledged the kids in the back seat. Just started talking.

The family looked around for a few minutes and left.

When I asked what happened, the salesperson said: "They were just looking."

No. They were watching. And what they saw told them everything they needed to know about what this experience would feel like.

## The Right Distance

Approach to a conversational distance — close enough to talk normally, far enough that the customer doesn't feel physically crowded. In most environments, that's roughly an arm's length to six feet. You're not trying to shake their hand from ten feet away, and you're not trying to get close enough that they take a step back.

Read the customer as you approach. Some people open up physically as you get closer — they turn toward you, they make eye contact, they signal they're ready to engage. Others hold their body language slightly closed, which means they need a little more space and a little more time before they're comfortable. Give it to them. A professional reads what's happening and adjusts. An amateur follows the script regardless of what the person in front of them is doing.

## Never Two at Once

One salesperson approaches. Not two. If your colleague got up at the same time you did, one of you turns around. A customer being approached by two salespeople at once doesn't feel like they're being well cared for. They feel like they're being cornered. One person handles the greeting. One person owns the relationship from this moment forward. That's the professional standard.

---

**"The Rule: The approach tells the customer what the rest of the experience will feel like. Walk with purpose, not panic. Every step is part of the greeting."**

---

CHAPTER FIVE

# The Introduction

Your Name, Your Question, and Why Simplicity Wins

You've acknowledged the customer. You've approached at the right pace, with the right body language, at a comfortable distance. Now you open your mouth. And this is where many salespeople overcomplicate something that, at its core, is one of the simplest parts of the entire process.

The introduction is not a performance. It is not a carefully crafted opening line designed to disarm the customer with cleverness. It is not a script that sounds like you rehearsed it in the mirror. The customer doesn't want a performance. They want a person. And the simplest, most direct way for you to show up as a person is almost always the strongest opening you can make.

## Keep It Simple

The structure is three parts. A greeting. Your name. A service-oriented question.

"Hi, my name is [your name]. Welcome to [dealership or business name]. How can I help you today?"

That's it. Friendly, clear, professional. Notice what's not in there. "What are you here to buy today?" is not in there. "What are you looking for?" is not in there. "Is there something specific I can show you?" is not in there. Those questions, however well-intentioned, put the customer on the defensive before you've earned the right to ask them anything. You haven't built a single ounce of trust yet. You're a stranger. And strangers asking people to commit to a buying intention before they've had a chance to breathe is exactly the dynamic that gives this profession its reputation.

"How can I help you today?" is different. It's open. It's service-oriented. It positions you as someone who is there for them rather than someone trying to extract something from them. And crucially, the answer to that question tells you everything you need to know to begin. The customer gave it to you voluntarily, without pressure, because you asked in a way that didn't feel like a trap.

## Your Name Matters

Say your name clearly. Not quickly, not as an afterthought, not mumbled in between the greeting and the question. Your name is the thing that makes this interaction personal rather than transactional. It creates a small but real sense of accountability — this specific person is taking responsibility for helping you. That matters to customers more than most salespeople realize.

You'd be surprised how many salespeople rush or skip the name entirely. They say "Hi, how can I help you?" and jump straight into the conversation, leaving the customer unsure who they're talking to. Later, if they want to come back and ask for you specifically—which is exactly what you want—they can't. They'll say, "I talked to someone, but I don't remember their name." And someone else will take the up.

Say your name. Say it distinctly. Make it easy to remember.

## Do Not Hand Over Your Business Card

This one runs counter to what some training programs teach, so let me be direct about it. Do not hand your business card to a customer during the meet-and-greet.

At the moment of introduction, you are a stranger. You have not done anything for this customer. You have not listened to them, helped them, shown them anything, or given them any reason to want to remember who you are. Handing them a card at that stage is presumptuous. At best, it reads as forward. At worst, it reads as aggressive — like you're trying to stake your claim on them before you've provided a single thing of value.

You know what happens to most of those cards? I've picked hundreds of them up off the ground on lots and in showrooms over the years. They go in a pocket and end up in the trash.

Give your card after you've provided value. After you've listened to them, helped them find something, and done something that gives them a reason to want to call you specifically. That's when a card means something. That's when they'll keep it.

### The Tone of the Question

How you ask "How can I help you today?" matters almost as much as the words themselves. Asked with genuine interest and calm confidence, it opens the door to a real conversation. Asked in a flat, automatic tone — the tone of someone going through the motions for the four hundredth time this month — it lands like a formality the customer has to get past before the real interaction begins.

You already know this person doesn't want to be here. Most of them came in braced for exactly the kind of interaction they've heard stories about. When your opening question sounds like you actually want to know the answer — like you're genuinely curious about what brought them in and genuinely interested in helping them get it — that is the first small moment of trust. And in this business, trust is built in small moments.

---

**"The Rule: The introduction is not where you sell yourself. It is where you open the door. Keep it simple, say your name, and ask how you can help. Everything else comes after."**

---

CHAPTER SIX

# Acknowledge Everyone in the Group

The Fastest Way to Lose a Deal Before It Starts

Here is one of the most reliable ways to kill a deal in the first two minutes without saying a single wrong thing to the person you're talking to.

Ignore everyone else.

A husband and wife walk in together. The husband speaks first, so the salesperson talks to the husband. For twenty minutes, the salesperson talks to the husband — shows him features, answers his questions, walks him through options — while the wife stands nearby feeling invisible. And then, at some point, the wife looks at her husband and says something like "I'm ready to go." And they leave. And the salesperson is standing there saying, "They were just looking."

They were not just looking. The wife was making a decision. And the decision she made was that she didn't feel respected, so her family wouldn't spend their money there.

I have watched this happen more times than I can count. It is one of the most preventable and most expensive mistakes in the business.

## Everyone in the Group Is a Decision Maker

When a customer walks in with other people — a spouse, a partner, a parent, an adult child, a friend who came along for moral support — every single one of those people influences what happens. You may not yet know who the primary decision-maker is. You almost certainly don't know who the veto power belongs to. So the professional move is simple: greet everyone.

Not in a performative way. Not with a production that makes it obvious you're running through a checklist. Just naturally, with the same warmth and attention you'd give the person who spoke first. Turn to the person who hasn't introduced themselves and say, "And who do I have the pleasure of meeting?" or simply "And you are?" Two seconds. It costs nothing. And it communicates something enormously important: you matter in this conversation too.

### The Truth About Spouses

In most cases, when a couple walks in together, the woman is the real decision maker. That's not a generalization — that's thirty-five years of experience. The man might drive the initial conversation. He might be the one who came in with the specific vehicle in mind and the strong opinions about trim levels and towing capacity. But if she's not comfortable, if she doesn't feel respected, if she goes home with a bad feeling about the experience — there is no deal. Happy wife, happy life is not just a saying. In this business, it is a closing strategy.

So when a couple walks in, and the husband introduces himself but doesn't introduce his wife, you turn to her. You make eye contact. You ask her name. You direct some of your questions to her. You make sure she knows that her opinion in this conversation carries weight — because it does — and that the professional acknowledges that from the beginning.

### Don't Forget the Children

If there are children in the group, acknowledge them too. You don't need to make a production of it. A simple "Hey, what's your name?" to a child who's old enough to answer, or even just a smile and a wave at a younger one, does something measurable. Parents notice when someone treats their kids with basic warmth and respect. It creates goodwill faster than almost anything else you can do in the first two minutes. And goodwill is exactly what you need the parents to feel before you ask them to spend a significant amount of money.

## Always Ask If They're Here to See Anyone

Before you go any further with the greeting, ask: "Have you been in before, or is this your first time visiting us?"

This is non-negotiable, and it has to happen early. If a customer spoke to someone else last week, or came in three days ago and spent an hour with another salesperson, and you take the up without asking, you are about to waste thirty minutes of everyone's time. At some point, they'll mention it. At that point, you have to turn them over to whoever they originally spoke with. You've done the work, and you get nothing.

Ask early. Ask cleanly. "Have you been in before, or is this your first time?" If they have a relationship with someone else at the store, find that person and make the introduction professionally. That's the right move for the customer, for your colleague, and for the dealership. It also tells the customer something important about the kind of organization they're dealing with.

### FROM THE FLOOR

I walked into a well-known furniture store looking for a recliner. Three salespeople were sitting on a showroom couch. I could hear them, in earshot, debating whose turn it was to help me — like I was interrupting something important.

The one who drew the short straw walked over and said, without a greeting, without a smile, without even telling me his name: "What are you here to buy today?"

My answer was immediate. "Nothing. I'm just looking. I'll let you know if I need help."

And I meant it. They'd lost me in the first five seconds. I didn't buy a thing. Went somewhere else.

The irony is that I walked in ready to buy. I knew what I wanted. All they had to do was make me feel like a person, not a chore, and the sale was theirs. Instead, they spent more energy arguing over whose turn it was than they spent on the customer standing in front of them. That's not a sales problem. That's a culture problem. And it starts with how the team treats the greeting.

---

**"The Rule: Greet every person in the group, every single time. The one you ignore is almost always the one who decides."**

---

CHAPTER SEVEN

# Common Mistakes That Kill the Greeting

What Goes Wrong in the First Thirty Seconds and How to Fix It

Most lost customers aren't lost at the close. They're lost in the first thirty seconds. Not because the salesperson was dishonest or incompetent or said something egregiously wrong. But because small habits — habits that feel harmless or even natural in the moment — communicate the wrong thing at the worst possible time.

These are the mistakes I've seen cost salespeople deals over 35 years in this business. They're common. They're fixable. And they're almost always invisible to the person making them.

## The Phone Problem

Walking up to a customer while holding your phone is one of the most damaging things you can do during the meet-and-greet, and it has become more common with every passing year. It doesn't matter if you were just checking the time. It doesn't matter if you just put it in your pocket as you stand up. The customer who saw you looking at it ten seconds before you approached them has already received the message: something on that screen was more interesting than they were.

The phone goes away before you approach. Not as you approach. Before. If that means putting it on your desk or leaving it in your pocket before you even stand up, that's what it means. A customer who has your undivided attention from the moment you start moving toward them feels it. It's a small thing that communicates a large idea.

## Speaking Too Fast or Too Softly

Nerves show up in the voice before almost anywhere else. When salespeople are anxious — especially newer ones — they rush. They want to get through the greeting quickly, past the awkward part, and into the conversation. So they talk faster than the customer can comfortably process. Instead of sounding confident, they sound anxious. An anxious salesperson does not inspire customer confidence.

The same problem appears in the other direction. Some people speak so softly in the greeting that the customer has to lean in and ask them to repeat themselves. That's an uncomfortable way to start a conversation, and it introduces a small but real friction right at the moment you most need everything to feel easy.

The fix for both is the same: slow down. A calm, clear, steady voice at a normal conversational volume communicates control. It tells the customer that you are comfortable in this environment and comfortable with this process. That comfort is contagious. When you're steady, they become steadier too.

## The Bad Attitude Carried Forward

Customer-facing work is emotionally demanding. Not every interaction goes well. Some customers are impatient. Some are rude. Some situations are simply frustrating. And one of the most common — and most damaging — mistakes in this business is carrying the emotional residue of one bad interaction into the next one.

A salesperson who just had a deal fall apart, dealt with a difficult customer, or got chewed out by a manager is carrying something. If they walk straight from that interaction to the next customer without resetting, the new customer absorbs it. Their tone is flat. Their smile is forced or absent. Their body language is tight. And the new customer, who did nothing to cause any of it, starts their experience on the wrong foot because of something they had nothing to do with.

The reset is not complicated. It doesn't require meditation or a lengthy break. It requires thirty seconds and a deliberate decision. Take a breath. Adjust your posture. Clear the last interaction out of your head. Remind yourself that the person about to walk through that door deserves a fresh start — because they do. That thirty-second reset can completely change the tone of the next greeting and every interaction that follows.

## Over-Eagerness

The sprint from Chapter Four is the most dramatic version of this mistake, but over-eagerness shows up in subtler ways, too. It's the salesperson who asks a closing question before the customer has had a chance to look around. It's the salesperson who volunteers information the customer didn't ask for because the silence feels uncomfortable. It's the salesperson who follows the customer from display to display at a distance of four feet, hovering without saying anything useful.

Customers need room. They need a moment to settle into the environment before they're ready to engage fully. The salesperson who gives them that room — who acknowledges, approaches, introduces, and then eases back slightly while remaining available — is the one who gets invited into the conversation. The one who fills every silence with presence and pressure is the one the customer is trying to get away from.

## Ignoring the Group

Covered in the previous chapter, but worth repeating here as a mistake category because it is that expensive. The customer you're talking to is not the only customer in the room. Everyone who walked in with them is evaluating the experience. Everyone who walked in with them can end the deal. Greet everyone. Every time. Without exception.

## The Wrong Opening Question

"What are you looking to buy today?"

"What brings you in?"

"Is there something specific I can show you?"

None of these is a terrible line. But all of them put the customer in the position of having to declare an intention before they're ready. Customers don't always know exactly what they want when they walk in. More importantly, even when they do, they're not necessarily ready to tell a stranger within the first ten seconds of meeting them.

"How can I help you today?" is different in a way that matters. It's open. It's non-threatening. It allows the customer to answer in whatever terms feel natural to them. That answer — whatever form it takes — is the beginning of real information. It is the customer telling you, voluntarily, where to start. That's worth more than any clever opener you could use in its place.

---

**"The Rule: The mistakes that kill the greeting are almost always invisible to the person making them. Know what they are so you can catch them before the customer does."**

---

CHAPTER EIGHT

# When Customers Say I'm Just Looking

What It Really Means and What to Do About It

Every salesperson who has spent more than a week on the floor has heard it. You approach, you introduce yourself, you ask how you can help — and the customer looks at you with a practiced expression and says: "I'm just looking."

And almost every new salesperson hears that phrase as rejection.

It isn't. In my experience, it seldom is. Understanding what "I'm just looking" actually means — and how to respond to it in a way that keeps the door open rather than closing it — is one of the most practical skills in this entire book.

## What the Phrase Actually Means

"I'm just looking" is a protective response. It is what people say when they're not ready to be pressured. It's a shield they raise before they know whether they need it — because they've been conditioned, by years of experiences in retail and sales environments, to expect that the moment they express any level of interest, someone is going to push them toward a decision they're not ready to make.

What the phrase rarely means is: I have no interest in what you're selling, I'm not going to buy anything today, and I would like you to leave me alone permanently. That's the interpretation most salespeople put on it, and it causes them to either withdraw completely or push back harder — both of which are the wrong response.

What it usually means is: I'm not ready to engage with you fully yet. I need a few more minutes to get comfortable in this environment. I'm protecting myself against pressure I haven't experienced yet but am expecting.

That is very different from rejection. That is a customer telling you, in the most polite way they know, what they need in this moment. And what they need is space.

## The Right Response

The correct response to "I'm just looking" is pressure-free acknowledgment. Something like: "That's perfectly fine. Take your time. I'll be nearby if you see something you want to take a closer look at."

Then give them that moment. Don't follow them at four feet. Don't hover. Don't fill the silence with information they didn't ask for. Let them breathe. Let them move through the space on their own terms and in their own time.

What happens next, more often than most salespeople expect, is that the customer comes back to you. They have a question. They've found something they want to look at more closely. They've decided on their own that you seem approachable and would like some help. The wall came down not because you pushed it but because you didn't. And a customer who lowers their own guard is in a completely different headspace than one whose guard is forced down.

## Sometimes It's Not About You

Here's something worth knowing: sometimes the "I'm just looking" response has nothing to do with the salesperson. Nothing to do with the approach, the tone, the opening question — none of it. The customer walked in with that shield already raised, from a previous experience somewhere else, from a story a friend told them, from a general wariness about this type of environment that has been building for years.

And sometimes it is about you — but not in the way you think. Sometimes a customer clicks with one salesperson but not another for reasons that have nothing to do with skill. A different energy. A resemblance to someone they don't like. An accent that reminds them of something. These things are real, and they're outside your control.

I had a colleague early in my career who was getting nowhere with a woman at the used-car lot. He told me she was just looking, that something was off, that he couldn't figure out what the problem was.

I went out, introduced myself, and asked how I could help her.

She said, "Thank God you're here. I really want to buy this car. But that other guy looks exactly like my ex-husband, and I cannot stand the sight of him."

Nothing to do with the car. Nothing to do with the price. Nothing to do with the dealership. We test-drove the vehicle, worked out fair numbers, and she drove home happy. My colleague had done nothing wrong. But the customer needed someone else in front of her before she was going to open up.

Don't take "I'm just looking" personally. Don't take it as a verdict. Take it as information: something needs to change. Sometimes that's giving more space. Sometimes that's a different approach. And sometimes — occasionally — that's a different person entirely.

## What You Never Do

You never argue with it. You never counter it with a sales pitch. You never say "That's fine, but let me just show you this one thing really quick" — because that's not respecting the signal they just gave you. That's telling the customer that what they said doesn't matter to you. And a customer who feels unheard in the first minute of an interaction is not going to trust you with a significant financial decision in minute thirty.

Respect the phrase. Respond to it without pressure. Stay available without hovering. Let the customer come to you when they're ready. That patience — that willingness to let the customer set the pace — is one of the things that separates professionals who build long careers in this business from the ones who burn through customers and wonder why nothing ever sticks.

---

**"The Rule: 'I'm just looking" is not rejection. It is a request for space. Give it generously, stay available, and let the customer come to you on their own terms."**

---

CHAPTER NINE

# The Ten-Second Rule

Why First Impressions Form Faster Than You Think

Researchers who study first impressions have consistently found that people form initial judgments of a new person within seconds — sometimes as few as two or three. In a sales environment, I give it ten seconds to be generous. In those ten seconds, the customer has already decided several things about you.

Are you confident or uncertain? Are you approachable or threatening? Do you look like someone who can help them, or someone who will make this harder than it needs to be? You don't get a second chance at those ten seconds. They happen whether you're ready or not. The only question is whether you've decided what those ten seconds are going to communicate.

## What Happens in Ten Seconds

The human brain processes nonverbal information extraordinarily fast. Long before the conscious mind has formed a complete thought about the person it's looking at, the subconscious has already run through a rapid-fire evaluation: safe or unsafe, competent or incompetent, interested or indifferent, worth engaging or worth avoiding.

This is not a flaw in how people think. It is an efficiency mechanism. In most social contexts, it works well. In a sales context, it means that everything you project in the first ten seconds — posture, expression, pace, energy, awareness — gets processed and evaluated before you've had the chance to say a single word.

A customer who sees a composed, attentive professional moving toward them with calm purpose feels something shift before the greeting even begins. A customer who sees a distracted, slouching, phone-occupied salesperson scrambling to their feet when they notice someone on the lot feels something shift, too — in the opposite direction. The ten-second rule doesn't reward effort. It rewards preparation.

## The Clock Starts Before You're Ready

Here is the uncomfortable implication of all this: you don't get to decide when the clock starts. The customer decides. The moment they see you — from their car in the parking lot, through the glass front of the showroom, as they walk through the door — the evaluation has begun. Not when you look up. Not when you decide to engage. When they see you.

That means your window to make a strong first impression is not the moment you open your mouth. It is the moment you become visible. Everything from your posture to the expression on your face to whether you're holding your phone is part of the first impression, and all of it is being registered before you've consciously decided to be "on."

The professionals who understand this are always on. Not in a performative way — in a prepared way. They've made their choices about how they're showing up before the first customer of the day arrives. Their posture, their energy, their awareness — those are decisions they made before the lot opened, not adjustments they scramble to make when someone pulls in.

## You Cannot Recover the First Ten Seconds

This is the part that trips people up. There is a persistent belief in sales that a strong recovery can overcome a weak opening. And there is some truth to that — a skilled salesperson can often salvage an interaction that started poorly. But salvaging is harder than getting it right. It requires extra work, extra time, and extra trust-building to overcome a deficit that never needed to exist.

More importantly, some customers don't give you the chance to recover. They form their impression in those first ten seconds, decide the experience isn't worth their time, and either leave immediately or disengage emotionally while remaining physically present. The salesperson who spent forty-five minutes with a customer who "wasn't interested" may have actually spent forty-five minutes with a customer who, in the first ten seconds, decided they weren't interested and was simply being polite while they looked for an exit.

The ten-second rule is not a reason to be anxious. It is a reason to be prepared. The preparation is simple: know what you're communicating, make conscious choices about it, and be ready before the customer arrives. That's it. Do those three things consistently, and the first ten seconds take care of themselves.

## FROM THE FLOOR

One afternoon, I was at my desk when I noticed a customer pull into the lot in a taxi, which caught my attention immediately. Most customers drive themselves. This one stepped out of the cab and walked directly toward a specific vehicle, like he knew exactly what he was looking for.

I didn't rush. I stood up, walked out at a normal pace, and gave him a small wave as I approached. When I reached him, I introduced myself and told him I'd be happy to help if he had any questions.

He told me he'd just gotten off a flight and had come straight from the airport. His vehicle had been destroyed in a fire in the airport parking lot while he was traveling. He'd seen one of our ads and came directly to us. He knew which vehicle he wanted. He just needed to drive it and confirm it.

We took a short test drive. Came back. He asked how to make out the check.

That's it. Start to finish, maybe forty-five minutes. The deal was easy because the approach was right. No pressure, no assumptions, no rushing. Just a professional greeting and a willingness to follow the customer's lead.

Not every customer comes in ready to buy. But every customer deserves that same professional approach. You never know which one will be the taxi customer.

---

**"The Rule: You don't get to decide when the clock starts. Be ready before the customer sees you, because the ten seconds that matter most happen before you open your mouth."**

---

CHAPTER TEN

# The Perfect Meet and Greet Script

A Framework for Every Situation, in Your Own Voice

Almost every salesperson, at some point, wants to know exactly what to say. Give me the line. Tell me what to say. Just tell me the script.

I understand the impulse. When you're new or when you're in a situation you haven't encountered before, having a reliable structure to fall back on is genuinely useful. The problem isn't wanting a script. The problem is treating the script as a performance to deliver rather than a framework to internalize.

The salespeople who sound scripted are reading a script. The salespeople who have truly internalized one — who understand the purpose of every element and have translated it into their own natural voice — don't sound scripted at all. They sound like professionals who know exactly what they're doing. That's the goal. And it takes repetition, practice, and the willingness to put in the work until the structure is automatic.

## The Structure

A strong meet-and-greet script has five elements. They happen in sequence, every time, with every customer.

First, the acknowledgment. Before you reach them, let them know you've seen them. A nod, a wave, a brief "Welcome in, I'll be right with you" from a distance. This can be nonverbal depending on proximity.

Second, the approach. Move toward them at a calm, purposeful pace. Hands visible, head up, phone away.

Third, the introduction. Your name, the business name, said clearly and with intention.

Fourth, the service question. Open-ended, non-threatening, focused on helping rather than selling.

Fifth, the listen. Ask the question and then stop talking. Let the customer answer in their own words, at their own pace, without interruption.

That's the entire framework. Five steps. None of them is complicated. All of them are trainable.

## What It Sounds Like

In most environments, the script sounds something like this:

> "Hi, welcome in. My name is [your name] — how can I help you today?"

That's it. Friendly, clear, professional. For a couple or group, it expands slightly:

> "Hello, welcome in. My name is [your name]. Great to have you both here — how can I help you today?"

And if you haven't yet confirmed whether they're here to see someone specific, that question comes immediately after:

> "Have you been in before, or is this your first time visiting us?"

Simple. Clean. Leaves nothing important out.

## What the Script Is Not

The script is not a clever opener. It is not a disarming line designed to lower the customer's guard with wit. It is not a memorized speech that starts with the same three sentences every time, regardless of context. It is a structure — a reliable map that ensures you cover what needs to be covered without missing anything, delivered in your voice and adapted to the situation in front of you.

In a rainy parking lot, your opening acknowledges the weather. "Welcome in — glad you made it through that. I'm [name], how can I help you?" On a hot afternoon, maybe it's: "Hey, come on in out of this heat — I'm [name], good to see you." The structure is the same. The delivery adapts to what's real in the moment.

That flexibility — the ability to use the framework as a foundation while responding to what's actually happening in front of you — is what makes a script feel like a conversation rather than a performance.

## How to Make It Yours

Read the script structure. Understand what each element is trying to accomplish. Then put every word in your own voice.

Not different words that accomplish different things. Your words accomplish the same things. The acknowledgment still happens. The approach is still calm and purposeful. Your name is still said clearly. The service question is still open-ended and non-threatening. The customer still gets to answer without being interrupted. All of that stays the same. The vocabulary and rhythm become yours.

Then practice it. Out loud. Not in your head — out loud. Record yourself and listen back. Does it sound natural? Does it sound like you're reading? Ask a colleague to play the customer and run through it cold. Do it until you stop thinking about the structure and start having a conversation.

This process takes weeks, not days. But once it's there — once the framework is so deeply internalized that you're not consciously running through steps anymore — your meet and greet becomes noticeably stronger. More confident. More consistent. More effective.

## Consistency Is the Standard

Here is the thing about scripts that most salespeople resist and all top performers eventually accept: the script is not for the easy customers. It is for all of them.

When the customer is warm, engaged, and clearly ready to talk, the script is easy to follow. You don't need it — the conversation flows naturally, and everything happens without effort. The script earns its value with the guarded customer, the distracted customer, and the customer who walks in having already decided they don't want to be there. Those are the moments when structure keeps you on track. Those are the moments when a strong, consistent framework is the difference between a conversation that opens and one that closes before it begins.

The salesperson who follows the framework 90% of the time gets 90% results at best. The ones at the top do it right a hundred percent of the time — not because they're more talented, but because they understand that consistency is the skill. The script is what makes consistency possible.

---

**"The Rule: Don't read the script. Own it. Put it in your own voice, practice it until it's automatic, and use it every single time without exception."**

---

CHAPTER ELEVEN

# Body Language in the Meet and Greet

When What You Do Speaks Louder Than What You Say

Body language often speaks louder than words in the first moments of an interaction. This is not a metaphor. It is how human communication actually works. Customers are processing your posture, facial expression, movement, and eye contact simultaneously, and they do so faster than conscious thought.

That means a person can say all the right words while still creating the wrong impression. The tone is warm, but the posture is closed. The words are welcoming, but the eyes keep drifting to the phone. The greeting is correct, but the body language suggests this interaction is a distraction. Customers don't analyze this consciously. They feel it. And what they feel shapes what they do next.

Understanding your body language — and making deliberate choices about it — is not a soft skill. It is one of the most practical steps you can take to improve the effectiveness of your meet-and-greet.

## Posture

Posture is the first nonverbal signal a customer reads, and they read it from a distance. Standing upright with relaxed shoulders communicates confidence and readiness. It says: I am in control of this environment, I am comfortable here, and I am capable of helping you.

Slouching communicates the opposite. So does leaning against a wall, sitting with your body collapsed into a chair, or standing in a way that looks like you're waiting for something more interesting to happen. These postures communicate low energy, low interest, or low investment in the job — and customers register that impression before you've taken a step toward them.

The adjustment is simple and costs nothing: stand well. Head up, shoulders back and relaxed, weight balanced. That's not a military bearing. It's the natural posture of someone who is present and prepared. Check it periodically throughout your shift. It drifts when you're tired or distracted, and customers notice when it does.

## Eye Contact

Natural eye contact communicates attention and honesty. It tells the customer: you are the focus of my awareness right now. Nothing else is competing for it.

This does not mean staring. Intense, unbroken eye contact creates a different kind of discomfort — it feels confrontational rather than welcoming. What you're aiming for is the kind of calm, attentive awareness that characterizes a conversation between two people who are both genuinely present. You look at them. When it's natural to look away briefly, you do. But your eyes keep returning to them, not to the floor, not to your phone, not to whatever is happening across the room.

When you're talking to a group, eye contact travels. You include everyone. You look at the person speaking, but you also check in with the others — a glance that says: I know you're here too, and your presence matters in this conversation. That's not complicated. It's just attentive.

## Facial Expression

Your face sets the emotional tone of the interaction before a word is spoken. A relaxed, open expression — even a small, genuine smile — lowers tension. It communicates welcome. It tells the customer that their arrival was a good thing, not an inconvenience.

A blank expression, a furrowed brow, or the flat look of someone running on autopilot creates distance. The customer doesn't know why the person in front of them looks slightly displeased. They don't know you're tired, or that the last customer was difficult, or that you've been standing on your feet for six hours. What they know is how your face makes them feel. And if your face makes them feel like they interrupted something, that's where the interaction starts.

The practical solution is the same one I referenced earlier in this book: make a deliberate choice before the customer arrives. Decide what you're projecting. If you need thirty seconds between interactions to reset, take them. That reset is not a luxury. It is part of the job.

### Hands and Movement

What you do with your hands matters. Fidgeting — with keys, a pen, your phone, the hem of your jacket — communicates nervousness. It draws attention away from the conversation and toward the distraction. Keeping your hands relaxed and still, or using them in natural, open gestures when you're talking, communicates groundedness and calm.

Movement matters too, and we've covered pace in detail in the approach chapter. But it extends beyond the approach. How you move when you're walking alongside a customer to show them a vehicle, how you position yourself when you're both looking at something, how you carry yourself through the space — all of it is body language. All of it is communicating something about who you are and how seriously you take this work.

### Alignment

The most useful way to think about body language is alignment. Your body should reinforce the message your words are trying to convey. If you're saying "How can I help you?", your posture, expression, eye contact, and movement should all convey that you are genuinely ready and willing to help. When words and body language align, the greeting feels authentic. When they contradict each other — when the words are warm, but the body is closed — customers feel the disconnect even if they can't name it.

Alignment is not performance. It is not about manufacturing enthusiasm you don't feel or projecting an image that isn't real. It is about making sure that when you genuinely want to help a customer — which is the baseline expectation of the job — your body is communicating that clearly rather than accidentally contradicting it.

Pay attention to what your body is saying. Most people don't. The ones who do have a significant advantage in the first ten seconds, before the conversation has even started.

---

**"The Rule: Your words open the door. Your body language determines whether the customer walks through it. Make sure they're saying the same thing."**

---

CHAPTER TWELVE

# The Psychology of First Impressions

Understanding Why the Greeting Affects Everything That Follows

Everything in this book comes back to one thing: the person on the other side of the interaction.

The steps of the meet-and-greet, the acknowledgment, the approach, the introduction, the script, the body language — all of it is just a framework for guiding a human being through a moment most of them find uncomfortable. The techniques work because people work a certain way. And the more clearly you understand how people work, the more effective every other tool in this book becomes.

## They're Already Nervous

Before a customer sets foot on your lot or walks through your door, they've already had a conversation in their head about what this experience is going to be like. For most of them, that conversation includes at least some version of the following: this is going to be uncomfortable, someone is going to pressure me, I might get taken advantage of, and I'd rather be doing almost anything else.

They've heard the stories. Their parents told them. Their friends warned them. Some of them have had bad experiences themselves. They arrive with their guard already up, and the first thing you need to do — before any selling happens at all — is bring that guard down.

Everything we've covered in this book is designed to do exactly that. The calm approach, the pressure-free greeting, the service-oriented question, the willingness to give them space when they say they're just looking. All of it is architecture for comfort. When a customer feels comfortable, they open up. When they open up, they tell you what they need. When you know what they need, you can help them get it.

Start every interaction from the understanding that the customer walked in tense. Your first job is to change that.

## They Don't Know the Process

Here is something every professional eventually forgets, because it happens to everyone who gets good at their job: you forget what it feels like not to know what you know.

When you've been in this business for a while, the process feels completely natural. The steps of the sale, the meet and greet sequence, the way a conversation moves from greeting to qualification to vehicle selection — it's all automatic. You move through it the way you get dressed in the morning: without thinking, without explaining, without registering that any of it might feel unfamiliar or intimidating to someone experiencing it for the first time.

But to the customer walking through your door, none of it is automatic. The terminology is foreign. The process is opaque. They don't know what's coming next. They don't know what's negotiable and what isn't. They don't know whether they're being treated fairly or taken advantage of. They're navigating a situation that you find completely routine, and for them, it may happen once every five or ten years.

That gap — between what you know and what they know — is where most of the friction in customer interactions comes from. Not from bad intentions. Just from the professional's natural tendency to forget what it feels like to be on the other side.

Your job is to close that gap. Not by explaining every step in detail — that would be overwhelming. But by being transparent enough that nothing feels like a surprise. By communicating what's happening and what comes next. By treating the customer like someone who deserves to understand the process, not someone who needs to be managed through it.

## People Want to Be Led, Not Pressured

This is the most counterintuitive principle in the entire book, and it is worth taking the time to understand it clearly.

Customers don't want to be pressured. But they do want to be led. There is a significant difference between the two, and it matters enormously during the meet-and-greet.

Pressure means forcing someone toward a decision they're not ready to make, in a way that overrides their judgment. That feels bad and creates resistance. Leading means guiding someone through a process they don't fully understand, in a way that respects their judgment while giving them the direction they need. That feels like service. And most customers will follow a confident, trustworthy guide through an unfamiliar process — because that is exactly what they came for.

Your customer buys a vehicle every three to ten years. You do this every day. You are the expert in this room. When you act like one — when you're confident, organized, and clear about what comes next — customers feel safe following you. That is not manipulation. That is professional service.

The moment you forget that you're the professional and start deferring to the customer on process questions they don't know the answers to, the whole thing gets wobbly. Stay confident. Know what comes next. Lead the way. That's what they need from you, even if they couldn't articulate it themselves.

## People Remember How You Made Them Feel

Customers don't remember every detail of an interaction. They may forget the exact wording of a greeting. They may forget your name. But they remember how the experience made them feel. That emotional impression — formed largely in the first few minutes of the interaction — often determines whether they come back, refer others, and tell the story of their experience as a positive one or a cautionary tale.

Two businesses can sell the same product at the same price. The one that makes people feel respected, comfortable, and genuinely helped is almost always the one that wins long-term loyalty. The meet-and-greet is the first taste of that feeling. Get it right, and you're building something. Get it wrong, and you're spending the rest of the interaction trying to recover ground you never should have lost.

A simple way to check yourself: if I were the customer, how would this greeting make me feel? That question — asked honestly, before the customer arrives and in the middle of the interaction — keeps the focus where it belongs. On the person. On their experience. On the feeling they're going to carry out the door when the conversation is over.

---

**"The Rule: Understand the person before you try to help them. Everything in the meet and greet flows from that."**

---

CHAPTER THIRTEEN

# The Meet and Greet Checklist

Ten Things That Have to Happen Every Time

Checklists exist because human beings are inconsistent. Not because they're careless or uncommitted, but because consistency under pressure, over the course of a full shift, day after day, is genuinely hard. A checklist turns important habits into repeatable behavior. It removes the reliance on memory and mood and replaces it with a reliable standard.

The best professionals in any field use checklists. Pilots use them before every flight. Surgeons use them before every procedure. Not because they don't know what they're doing, but because the cost of missing something is high enough that they don't leave it to chance. The meet-and-greet is no different. The cost of missing an item on this list is a customer who walks out before you ever have a real conversation.

Run through this list mentally before every customer interaction. It takes less than ten seconds. It prevents most of the mistakes covered in this book.

✓ **Appearance is appropriate and professional for the environment.** — Clean clothes, groomed, ready to be seen before you step onto the floor.

✓ **Phone is in your pocket, not in your hand.** — Put it away before you approach, not as you approach.

✓ **You acknowledged the customer before approaching.** — A nod, a wave, or a brief verbal acknowledgment from a distance. They know they've been seen.

✓ **You approached at a normal walking pace.** — Not running. Not dragging. Purposeful and calm.

✓ **You introduced yourself by name.** — Said clearly, not mumbled, not rushed. Your name makes this interaction personal.

✓ **You asked how you can help — not what they're here to buy.** — Open, service-oriented, non-threatening. Let the customer answer in their own words.

✓ **You greeted every person in the group.** — Spouse, partner, parent, adult child, friend. Everyone who walks in gets acknowledged.

✓ **You asked whether they've been in before.** — Before you go any further. This protects your colleagues and the customer relationship.

✓ **You did not hand them a business card.** — Not yet. The card comes after you've provided value, not before.

✓ **You did not ask about payment, trade, or budget.** — Not in the meet and greet. That conversation comes after trust is established.

Ten items. None of them is complicated. All of them are trainable. The salespeople who do all ten, every time, consistently outperform the ones who do seven or eight and think that's close enough.

Close enough is not the standard. Every time is the standard.

## Using the Checklist in Training

This checklist is also a training tool. Managers who want to improve greeting quality on their team can observe interactions and track these ten items specifically. Not as a gotcha exercise, but as a coaching framework. Which items are being hit consistently? Which ones are being skipped? Where are the patterns?

Small corrections given early—slow down, put the phone away before you stand up, and acknowledge the wife—prevent bad habits from calcifying. A new salesperson who receives specific, observable feedback on these ten items will improve quickly. A new salesperson who is told to "do better" will not, because they have no idea what better looks like.

Great meet-and-greet behavior is not a personality trait. It is a skill. Skills are taught with clear standards, observable metrics, and specific feedback. This checklist provides all three.

## When You Miss One

You will miss items on this list. Everyone does. The goal is not perfection over the course of a career. The goal is a high enough rate of consistency that missing an item is the exception rather than the rule, and that when you do miss one, you notice it quickly enough to course-correct.

The professional habit is not to run a perfect checklist every time. It is to run a conscious one. To be aware enough of what needs to happen that you catch the gaps before the customer does. That awareness — the discipline of paying attention to your own process — is what separates the professionals who keep getting better from the ones who stop improving the moment they think they know what they're doing.

---

**"The Rule: Ten items. Every customer. Every time. Close enough is not the standard — every time is the standard."**

---

CONCLUSION

# The Professional Advantage

What Mastering This Skill Actually Gives You

In most businesses, people spend a great deal of time learning products, procedures, systems, and policies. All of that matters. But the human side of the interaction — the part that determines whether those skills ever get a chance to work — is the part that most training programs underinvest in.

If the customer relationship starts badly, even strong product knowledge may not be enough to restore lost trust. If the first impression communicates indifference, pressure, or disorganization, the customer's guard goes up and stays up. And a customer with their guard up is not a customer you can help.

The meet-and-greet is a professional advantage because it affects every other stage of the interaction. A strong greeting opens the door to better questions, better listening, better rapport, and better outcomes. It lowers tension and increases comfort. It tells the customer — before anything else has happened — that this experience will be different from what they were braced for.

## What You Now Know

You know that the meet and greet starts before you speak — that the impression forming in the customer's mind begins the moment they see you, not the moment you open your mouth.

You know what you're communicating before you say a word — posture, eye contact, expression, movement, and what you're holding all tell a story before the conversation begins.

You know how to acknowledge a customer in a way that tells them they've been seen without making them feel ambushed. You know how to approach with purpose rather than panic. You know how to introduce yourself clearly, ask the right question, and listen to the answer.

You know to greet everyone in the group — because the person you ignore is often the one who decides. You know the common mistakes that kill the greeting and how to catch them in yourself before they cost you. You know what "I'm just looking" really means and how to respond in a way that keeps the door open.

You understand the ten-second rule — that the clock starts when the customer sees you and that preparation is the only thing that controls those first seconds. You have a script structure you can make your own, practice until it sounds like a conversation rather than a performance. You understand body language as alignment, not performance.

You understand the psychology underneath all of it — why customers arrive tense, why they need to be led rather than pressured, why they remember how you made them feel long after they've forgotten everything else about the interaction.

And you have a ten-item checklist that runs through your head in less than ten seconds before every customer interaction, preventing most of the mistakes that cost salespeople deals every single day.

## What Comes Next

Knowing this material is the beginning, not the end. The professional advantage doesn't come from reading about the meet-and-greet. It comes from practicing it, refining it, and doing it consistently enough that it becomes part of how you work rather than something you have to think about.

Practice the acknowledgment until it's automatic. Practice the approach until it feels natural. Practice the introduction until your name comes out clearly every time without effort. Practice the script in your own voice until the structure disappears into the conversation and what's left is just you, talking to a person, helping them figure out what they need.

That level of consistency takes time. It takes repetition. It takes the willingness to receive feedback, notice your own gaps, and keep improving even when you're already performing at a high level. The professionals who stay at the top of this business never stop working on the fundamentals. They understand that mastery is not a destination. It is a practice.

### The Simplest Summary

Be ready before they arrive. Acknowledge them the moment you see them. Approach with purpose and calm. Say your name. Ask how you can help. Listen to the answer. Greet everyone in the group. Stay available without hovering. Give them space when they need it. Lead them forward when they're ready.

None of that is complicated. All of it is powerful. And the professionals who do it consistently — not most of the time, not when they feel like it, but every single time with every single customer — stand out in an industry where the baseline expectation is embarrassingly low.

The bar for exceeding expectations in this business is not high. You just have to clear it every time.

That is the professional advantage. And it starts the moment they see you.

— Bruce Huddleston

APPENDIX

# The Rules

Every chapter of this book ended with a rule. Here they are collected in one place — a quick-reference summary of the principles that matter most.

**Introduction** — "The Rule: The greeting is not where the sale is won. It is where the sale is lost. Everything else depends on getting this right."

**Chapter One** — "The Rule: The meet and greet starts the moment they see you. Be ready before they arrive, because once they do, the clock is already running."

**Chapter Two** — "The Rule: You are always communicating. The only question is whether you're doing it on purpose."

**Chapter Three** — "The Rule: Acknowledge every customer the moment they arrive. Not with pressure — with presence. Let them know they've been seen before you've said a word."

**Chapter Four** — "The Rule: The approach tells the customer what the rest of the experience will feel like. Walk with purpose, not panic. Every step is part of the greeting."

**Chapter Five** — "The Rule: The introduction is not where you sell yourself. It is where you open the door. Keep it simple, say your name, and ask how you can help. Everything else comes after."

**Chapter Six** — "The Rule: Greet every person in the group, every single time. The one you ignore is almost always the one who decides."

**Chapter Seven** — "The Rule: The mistakes that kill the greeting are almost always invisible to the person making them. Know what they are so you can catch them before the customer does."

**Chapter Eight** — "The Rule: 'I'm just looking" is not rejection. It is a request for space. Give it generously, stay available, and let the customer come to you on their own terms."

**Chapter Nine** — "The Rule: You don't get to decide when the clock starts. Be ready before the customer sees you, because the ten seconds that matter most happen before you open your mouth."

**Chapter Ten** — "The Rule: Don't read the script. Own it. Put it in your own voice, practice it until it's automatic, and use it every single time without exception."

**Chapter Eleven** — "The Rule: Your words open the door. Your body language determines whether the customer walks through it. Make sure they're saying the same thing."

**Chapter Twelve** — "The Rule: Understand the person before you try to help them. Everything in the meet and greet flows from that."

**Chapter Thirteen** — "The Rule: Ten items. Every customer. Every time. Close enough is not the standard — every time is the standard."

ALSO AVAILABLE

# By Bruce Huddleston

The Meet and Greet Playbook is part of The Car Sales Survival Series — a growing collection of focused training guides that go deep into one specific sales skill. Each book is designed to stand alone as a focused training resource or to be used as part of a complete system for onboarding and developing sales staff.

**The Complete Car Sales Survival Guide**

The No-BS Playbook for New Automotive Salespeople

## The Car Sales Survival Series

**The Meet and Greet Playbook**

How to Make Powerful First Impressions with Customers, Clients, and Guests

**The First 60 Seconds in Car Sales**

A Proven Meet and Greet System to Build Trust and Start More Conversations

**How to Handle "I'm Just Looking" in Car Sales**

A Simple System to Turn Brush-Offs into Productive Conversations

**Body Language in Car Sales**

How Posture, Eye Contact, and Presence Build Customer Trust

**Greeting Customers on the Lot**

How to Approach Buyers Without Pressure

**The Ten-Second Rule in Car Sales**

Why First Impressions Determine Whether Customers Stay or Leave

**The Car Sales Conversation Starter Guide**

How to Begin Natural Conversations That Lead to Sales

**Car Sales Confidence for New Salespeople**

How to Approach Customers Without Fear or Hesitation

**Common Car Sales Greeting Mistakes**

What Drives Customers Away in the First Minute

**The First Five Minutes With a Car Buyer**

How to Transition from Greeting to Conversation and Move Toward the Sale

# Coaching and Group Training

This book is a starting point. For salespeople and teams who want to go deeper — working through the material personally, applying it to their specific situation, and building the habits that make it stick — Bruce offers individual and group sessions through Life Guidance Consulting LLC.

## One-on-One Coaching

Individual sessions for salespeople at any stage of their career. Whether you're in your first ninety days on the floor or you've been selling for years and want to break through a plateau, one-on-one coaching gives you direct access to thirty-five years of real-world experience — applied specifically to your situation, your dealership, and your goals.

## Group Training Sessions

Group sessions for sales teams working through The Meet and Greet Playbook or the Professional Skills Training Series together. Ideal for dealership onboarding, team development, and ongoing skills training. Sessions are practical, direct, and built around real scenarios from the sales floor—not theory.

## Bulk Book Orders

Dealerships and sales organizations interested in using The Meet and Greet Playbook or the Professional Skills Training Series as structured onboarding or training materials can inquire about bulk pricing and customized packages.

**Life Guidance Consulting LLC**

www.lifeguidanceconsulting.com

bruce@lifeguidanceconsulting.com

For book orders and publishing inquiries:

www.bedrockheritagepublishing.com

info@bedrockheritagepublishing.com

# ABOUT THE AUTHOR

Bruce Huddleston spent thirty-five years in the automotive industry, working every level of the business from showroom floor salesperson to finance manager, sales manager, used car manager, and general manager. His career included new-car franchise dealerships, independent used-car operations, and a decade in buy-here, pay-here — giving him a breadth of experience that few in the industry can match.

He began as a high school dropout who needed a job and ended up discovering a profession. He ended as a veteran who had trained hundreds of salespeople, managed multiple departments, and built a reputation for straight talk in an industry that doesn't always reward it.

Since retiring, Bruce has opened a life coaching practice, assists his wife with her mental health therapy practice, and operates Bedrock Heritage Publishing, a division of Life Guidance Consulting LLC, where he writes practical guides for sales professionals across multiple industries.

The Complete Car Sales Survival Guide is his flagship work. Of The Car Sales Survival Series— a collection of focused training guides on specific sales skills — is built on the same foundation of real experience, honest insight, and zero tolerance for the kind of nonsense that gives sales a bad name.

He lives in Tyler, Texas.

www.lifeguidanceconsulting.com

www.bedrockheritagepublishing.com

www.carsalessurvivalseries.com

A QUICK FAVOR

# Did This Book Help You?

If The Meet and Greet Playbook gave you something useful a technique that clicked, a story that stuck, a habit that made your first thirty seconds noticeably stronger the single best thing you can do to help other salespeople find it is leave an honest review on Amazon.

It takes about two minutes. It makes a real difference to how the book gets discovered. And it helps the next salesperson who needs this information actually find it.

You can simply scan the QR code below

Thank you for reading.

https://www.amazon.com/review/create-review/?asin=1972179071

Bruce Huddleston

Made in the USA
Monee, IL
04 November 2026

67379641R00046